COMMON CORE SCIENCE

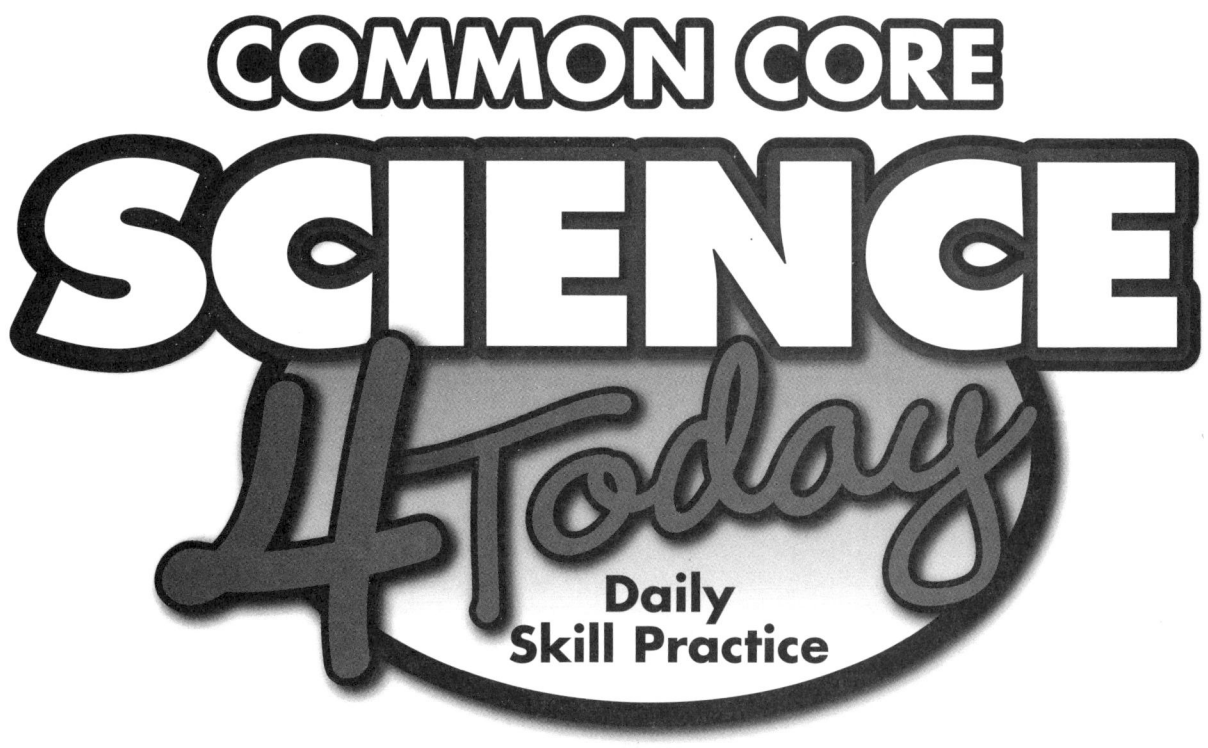

4 Today

Daily Skill Practice

Grade 1

Natalie Rompella

Carson-Dellosa Publishing, LLC
Greensboro, North Carolina

Credits

Content Editor: Elise Craver
Copy Editor: Karen Seberg

 Visit *carsondellosa.com* for correlations to Common Core, state, national, and
Canadian provincial standards.

Carson-Dellosa Publishing, LLC
PO Box 35665
Greensboro, NC 27425 USA
carsondellosa.com

ISBN 978-1-4838-1124-6
01-135141151

Table of Contents

Common Core Science 4 Today is a perfect supplement to any classroom science curriculum. Students' science skills will grow as they support their knowledge of science topics with a variety of engaging activities.

This book covers 40 weeks of daily practice. You may choose to work on the topics in the order presented or pick the topic that best reinforces your science curriculum for that week. During the course of four days, students take about 10 minutes to complete questions and activities focused on a science topic. On the fifth day, students complete a short assessment on the topic.

Various skills and concepts in math and English language arts are reinforced throughout the book through activities that align to the Common Core State Standards. Due to the nature of the Speaking and Listening standards, classroom time constraints, and the format of the book, students may be asked to record verbal responses. You may wish to have students share their answers as time allows. To view these standards, please see the Common Core State Standards Alignment Matrix on pages 5–8.

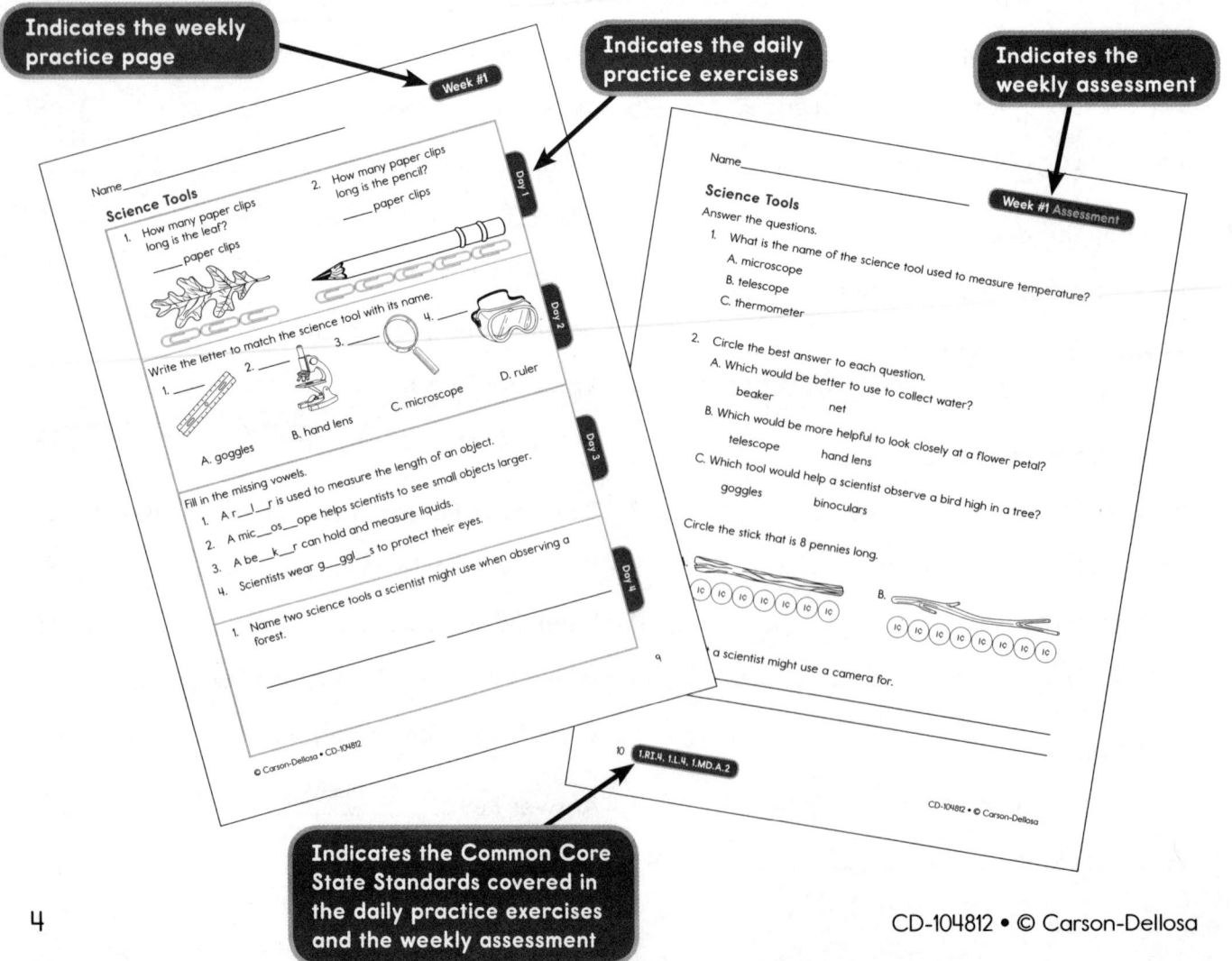

Indicates the weekly practice page

Indicates the daily practice exercises

Indicates the weekly assessment

Indicates the Common Core State Standards covered in the daily practice exercises and the weekly assessment

English Language Arts

STANDARD	W1	W2	W3	W4	W5	W6	W7	W8	W9	W10	W11	W12	W13	W14	W15	W16	W17	W18	W19	W20
1.RI.1									●				●							
1.RI.2								●												
1.RI.3							●			●			●							●
1.RI.4	●			●	●	●		●	●		●	●	●	●		●	●	●		
1.RI.5																				
1.RI.6																				
1.RI.7										●					●	●				●
1.RI.8																				
1.RI.9																				
1.RI.10																				
1.RF.1								●												
1.RF.3							●							●						
1.RF.4				●									●							
1.W.1																				
1.W.2						●														
1.W.3																	●			
1.W.5																				
1.W.6																				
1.W.7																				
1.W.8																				
1.SL.1																	●			
1.SL.2																				
1.SL.3																				
1.SL.4								●	●		●									
1.SL.5						●						●	●	●	●	●	●			
1.SL.6						●	●										●		●	
1.L.1		●		●		●	●	●			●						●			
1.L.2											●						●			
1.L.3																				
1.L.4	●				●	●		●				●		●			●	●	●	
1.L.5		●	●		●		●		●		●	●	●	●	●	●		●	●	●
1.L.6													●							

W = Week

English Language Arts

STANDARD	W21	W22	W23	W24	W25	W26	W27	W28	W29	W30	W31	W32	W33	W34	W35	W36	W37	W38	W39	W40
1.RI.1	●		●		●															
1.RI.2																				
1.RI.3		●		●	●	●	●	●	●	●		●			●			●		●
1.RI.4		●				●		●	●	●	●	●		●				●	●	●
1.RI.5																				
1.RI.6																				
1.RI.7	●					●					●								●	
1.RI.8																				
1.RI.9																				
1.RI.10																				
1.RF.1																				
1.RF.3					●					●			●	●	●		●		●	
1.RF.4																				
1.W.1																●				
1.W.2															●		●	●	●	●
1.W.3														●						
1.W.5																				
1.W.6																				
1.W.7																				
1.W.8														●			●			
1.SL.1										●										
1.SL.2																				
1.SL.3																				
1.SL.4		●		●	●					●					●			●		●
1.SL.5	●		●					●		●						●	●	●		
1.SL.6				●					●						●	●				
1.L.1													●	●	●	●	●	●	●	●
1.L.2							●		●					●		●				
1.L.3																				
1.L.4	●	●	●		●			●	●		●			●			●		●	
1.L.5	●	●	●		●			●	●			●	●					●	●	●
1.L.6																				

W = Week

CD-104812 • © Carson-Dellosa

Common Core State Standards Alignment Matrix

Math

STANDARD	W1	W2	W3	W4	W5	W6	W7	W8	W9	W10	W11	W12	W13	W14	W15	W16	W17	W18	W19	W20
1.OA.A.1			●						●	●								●		
1.OA.A.2		●	●																	
1.OA.B.3										●										
1.OA.B.4																				
1.OA.C.5									●				●							
1.OA.C.6				●																
1.OA.D.7																				
1.OA.D.8																				
1.NBT.A.1										●										
1.NBT.B.2																				
1.NBT.B.3			●																	
1.NBT.C.4				●																
1.NBT.C.5																				
1.NBT.C.6																●				
1.MD.A.1			●																	
1.MD.A.2	●		●																	
1.MD.B.3																				
1.MD.C.4		●		●					●		●		●			●		●		
1.G.A.1											●				●					
1.G.A.2																				
1.G.A.3																				

W = Week

Math

STANDARD	W21	W22	W23	W24	W25	W26	W27	W28	W29	W30	W31	W32	W33	W34	W35	W36	W37	W38	W39	W40
1.OA.A.1	●	●		●	●							●		●			●		●	
1.OA.A.2	●	●					●							●						
1.OA.B.3																				
1.OA.B.4		●				●														
1.OA.C.5																				
1.OA.C.6																				
1.OA.D.7																				
1.OA.D.8																				
1.NBT.A.1																				
1.NBT.B.2																				
1.NBT.B.3							●													
1.NBT.C.4												●								
1.NBT.C.5																				
1.NBT.C.6																				
1.MD.A.1											●									
1.MD.A.2																				
1.MD.B.3								●		●										
1.MD.C.4				●	●						●	●								
1.G.A.1											●									
1.G.A.2											●					●				
1.G.A.3																				

W = Week

Name_____

Science Tools

1. How many paper clips long is the leaf?

   _____ paper clips

2. How many paper clips long is the pencil?

   _____ paper clips

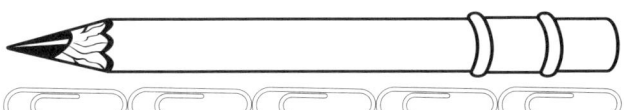

Write the letter to match the science tool with its name.

1. _____ 2. _____ 3. _____ 4. _____

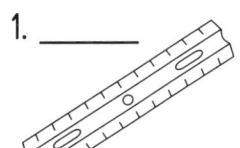 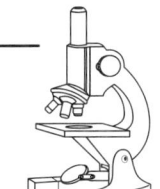  

A. goggles   B. hand lens   C. microscope   D. ruler

Fill in the missing vowels.

1. A r__l__r is used to measure the length of an object.

2. A mic__os__ope helps scientists to see small objects larger.

3. A be__k__r can hold and measure liquids.

4. Scientists wear g__ggl__s to protect their eyes.

1. Name two science tools a scientist might use when observing a forest.

   _____ _____

Science Tools

Answer the questions.

1. What is the name of the science tool used to measure temperature?

   A. microscope

   B. telescope

   C. thermometer

2. Circle the best answer to each question.

   A. Which would be better to use to collect water?

        beaker   net

   B. Which would be more helpful to look closely at a flower petal?

        telescope   hand lens

   C. Which tool would help a scientist observe a bird high in a tree?

        goggles   binoculars

3. Circle the stick that is 8 pennies long.

   A. 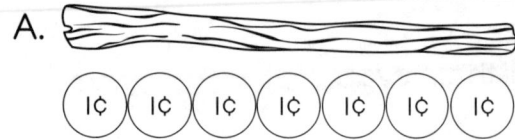   B.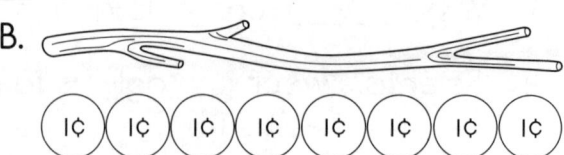

4. Tell what a scientist might use a camera for.

   _____

   _____

Name_____

Sorting and Classification

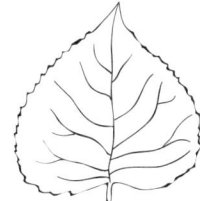   

1. Name two ways these leaves could be sorted.

_____   _____

1. Tia has sorted the animals into two groups. Fill in the top of the chart to show how she sorted them.

butterfly	cat
owl	worm
bee	snake
eagle	alligator

A biologist observed the plants and animals in and around a nearby pond. She counted what she saw.

lily pads: 4	frogs: 3	cattails: 4	turtles: 5
birds: 6	wildflowers: 9	dragonflies: 2	

1. How many total plants did she find? _____ plants

2. How many total animals did she find? _____ animals

1. Fill in the chart with six items you can observe in your classroom.

Heavy	Light

Sorting and Classification

Answer the questions.

1. Tell about something you have sorted into categories at home.

   _____

   _____

2. Draw a chart. Then, sort the animals into categories.

zebra

lion

fish

tiger

bat

bird

3. How did you sort them?

   _____

   _____

1.L.1, 1.L.5, 1.OA.A.2, 1.MD.C.4

Measurement

1. How many centimeters long is the stick?

_____ cm

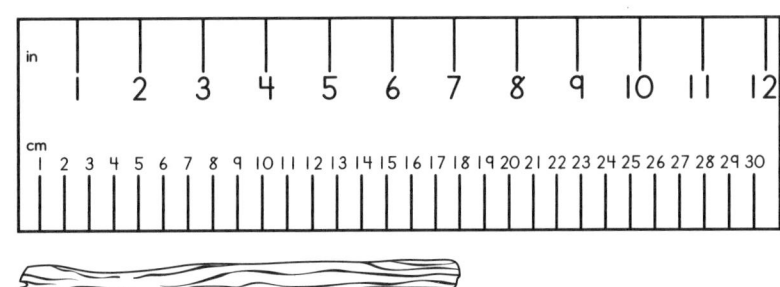

Circle the best answer.

1. amount of time from 6:00 am to 6:00 the next morning

   24 hours   24 seconds   24 minutes

2. time it takes to wash your hands

   30 seconds   30 minutes   30 hours

3. how often the garbage truck comes to your house

   every day   every week   every month

Use the chart to fill in the blanks.

> 100 centimeters = 1 meter
> 1,000 milligrams = 1 gram

1. A paper clip has a mass of about 1,000 milligrams, or _____ gram.

2. A baseball bat is about 100 centimeters, or 1 _____, long.

Look at the balance scale.

1. Which is heavier: the apple or the counters?

   _____

2. How do you know? _____

   _____

Measurement

Answer the questions.

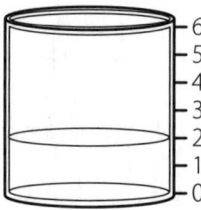 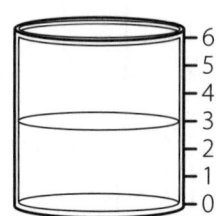 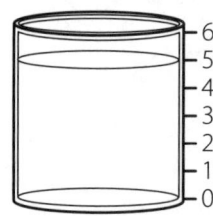

1. If you poured the 3 beakers of water into 1 larger beaker, how much water would you have? _____ mL

2. Write the name of the insects in order from smallest to largest.

    luna moth = 80 mm   flea = 2 mm

    bumblebee = 12 mm   grasshopper = 40 mm

    _____ , _____ , _____ , _____

3. Write the name of the science tool that would be needed.

    A. to measure length _____

    B. to measure weight _____

    C. to measure temperature _____

    D. to measure time _____

4. Fill in the blanks with the correct unit of measure.

Celsius	centimeters	grams	seconds

    A. A banana has a mass of about 140 _____.

    B. It takes about 15 _____ to say the alphabet.

    C. A pencil is about 17 _____ long.

    D. The temperature inside the classroom is about
       20° _____.

    CD-104812 • © Carson-Dellosa

Science Process Skills

Day 1

1. How might a scientist use math?

_____

_____

_____

_____

Day 2

1. What are some steps to follow when doing an experiment about which objects sink and float?

_____

_____

_____

_____

Day 3

Olivia and Mason each built a bridge with toothpicks. They then tested which bridge was stronger by placing coins on top. Olivia's bridge held 29 pennies before it began to break. Mason's bridge held 24 nickels before it began to break.

1. What makes it hard to tell whose bridge was stronger?

_____

_____

Day 4

Henry experimented to see if salt water or freshwater freezes faster.

1. What question is Henry trying to answer?

_____

2. What equipment might he need? _____

3. What do you think the results of Henry's experiment will be? Why?

_____

Science Process Skills

Answer the questions.

Marisa wanted to know which type of birdseed the birds outside her window liked best. Here is her experiment.

My Plan

1. Buy birdseed.

2. Sort the seeds.

3. Count out 10 seeds of each seed type. Put out dishes, each with one type of seed.

4. After 3 hours, count the seeds left in each dish.

My Prediction

I think the birds will like striped sunflower seeds best.

My Results

Type of Seed	Seeds Left	Seeds Eaten
black oil sunflower seeds	2	8
striped sunflower seeds	5	5
safflower seeds	10	0
white millet	7	3

1. Which type of seed did the birds eat most? _____

2. Which type of seed did the birds eat least? _____

3. Was Marisa's prediction correct? _____

4. Marisa used 2 different types of sunflower seeds. How many total sunflower seeds were leftover? _____

5. How many total seeds were eaten? _____

1.RI.4, 1.RF.4, 1.L.1, 1.OA.C.6, 1.NBT.C.4, 1.MD.C.4   CD-104812 • © Carson-Dellosa

Science Safety

1. Why do scientists wear safety goggles?

   A. to magnify what they are viewing

   B. to protect their eyes

   C. to make things look 3-D

1. Why is it a good idea to wash your hands after doing a science experiment?

   _____

   _____

   _____

   _____

Draw a line to match the safety equipment to its use.

1.	fire extinguisher	A.	to clean up spills
2.	latex gloves	B.	to put out fires
3.	paper towels	C.	to protect the hands
4.	goggles	D.	to protect the eyes

1. Circle each word that means something is dangerous.

   recycle   hazard   poison

   flammable   natural

Science Safety

Answer the questions.

1. Why should you clean up spills right away?

   _____

   _____

   _____

Complete the safety rules. Unscramble the letters to fill in the missing words.

2. Be sure to tie back long _____ (airh).

3. Never eat _____ (ofod) while doing science experiments.

4. Do not _____ (ridnk) liquids you are working with in science experiments.

5. What are some safety rules you follow in your classroom?

   _____

   _____

6. Draw a picture of a student not following three safety rules. Below the picture, tell what the student should have done instead.

   ┌───┐
   │   │
   │   │
   │   │
   │   │
   │   │
   │   │
   │   │
   └───┘

   _____

   _____

1.RI.4, 1.L.4, 1.L.5

The Five Senses

Complete the sentences. Unscramble the letters to fill in the missing words.

1. You use your nose to _____ (lmsel).

2. You use your eyes to _____ (ese)

3. You use your ears to _____ (erah).

4. You use your tongue to _____ (satet).

5. You use your hands to _____ (cuhto).

Day 1

1. What is your favorite food? _____

2. Is it bitter, salty, sweet, or sour? _____

3. What is your least favorite food? _____

4. Is it bitter, salty, sweet, or sour? _____

Day 2

1. Name something that feels hot. _____

2. Name something that feels cold. _____

3. Name something that feels prickly. _____

4. Name something that feels smooth. _____

Day 3

1. Tell how a scientist could use her senses to observe the woods.

_____

_____

_____

_____

_____

Day 4

The Five Senses

Answer the questions.

1. Think about your favorite fruit. How would you describe it to a friend? Use each of your senses to help describe it. Then, draw a picture to show your thoughts.

   _____

   _____

   _____

   _____

2. Think about your least favorite fruit. How would you describe it to a friend? Use each of your senses to help describe it. Then, draw a picture to show your thoughts.

   _____

   _____

   _____

   _____

3. Fill in the chart with words you used in your two paragraphs.

Words Describing Your Favorite Fruit	Words Describing Your Least Favorite Fruit

1.RI.4, 1.W.2, 1.SL.5, 1.SL.6, 1.L.1, 1.L.4

Scientific Inquiry

1. You want to observe how quickly the moon travels across the sky. Circle the tools that might help you.

   magnifying glass   telescope   compass

   pencil   camera   notebook

Fill in the blanks with letters to complete the question words.

1. w__o   2. w___ ___t   3. wh__r__

4. w__e__   5. h___ ___   6. w__y

7. wh__ch

1. Your school plans to start a school garden. Tell two questions you might have before you start.

   _____

   _____

   _____

   _____

Marcus wanted to learn more about the insects near his house. After coming up with a question, he observed the insects. Here are his notes.

Insect	Time	What It Ate
grasshopper	10:00 am	leaf
butterfly	10:05 am	nothing
grasshopper	10:30 am	nothing

1. What question do you think Marcus asked?

   _____

   _____

Scientific Inquiry

Answer the questions.

You want to learn more about the sun.

1. What questions might you ask?

   _____

   _____

   _____

   _____

2. How could you find your answers?

   _____

   _____

   _____

   _____

3. What are some ways you could organize your answers?

   _____

   _____

   _____

   _____

Fact vs. Opinion

Write **F** if something is a fact and **O** if something is an opinion.

1. _____ Apples taste good.

2. _____ Apples are a fruit.

3. _____ Apples grow on trees.

4. _____ An apple tree is pretty.

Day 1

1. Write a fact about bananas.

   _____

2. Write an opinion about bananas.

   _____

Day 2

Match the word with the definition.

1. How someone feels about something   A.   fact

2. Something that is true and can be proven   B.   opinion

Day 3

1. Mario is writing a paragraph about ladybugs. Draw a line through any sentences that are opinions.

Ladybugs are insects. They are pretty. Ladybugs eat aphids. Aphids are bad. They eat crops. Ladybugs have two sets of wings. They lift their top wings and use their thinner wings beneath to fly. They fly fast.

Day 4

Fact vs. Opinion

Answer the questions.

Write **F** if something is a fact and **O** if something is an opinion.

1. _____ Bananas are a kind of fruit.

2. _____ Oranges are the best-tasting fruit.

3. _____ Juice does not taste as good as fruit.

4. _____ Some animals eat fruit.

5. Write a paragraph on a topic you know a lot about. Include facts and details.

_____

_____

_____

_____

_____

_____

_____

_____

_____

_____

_____

_____

6. Write two opinion sentences about the same topic.

_____

_____

Observation and Inference

1. Look around you. Write down four observations you have made.

   _____

   _____

   _____

   _____

1. Circle the animals hidden in the picture.

Read through Katie's journal entry to answer the questions.

   Today, it was sunny out. I quickly put on my swimsuit and went outside. And there, on the sand, was a sea star! Then, I saw two pretty shells. When I went to pick one up, the water washed it away before I could grab it.

1. Where do you think Katie is? _____

2. What are some clues? _____

3. What do you think the temperature is? Why? _____

   _____

Match each set of observations with where it would likely be found in a school.

1. classroom

2. lunchroom

3. music room

A. Students are drinking milk. There are napkins and forks.

B. Students are holding pencils. There are desks. A teacher is writing on the board.

C. Students are singing. The teacher is playing a song. There is a piano.

Observation and Inference

Use the graph to answer the questions.

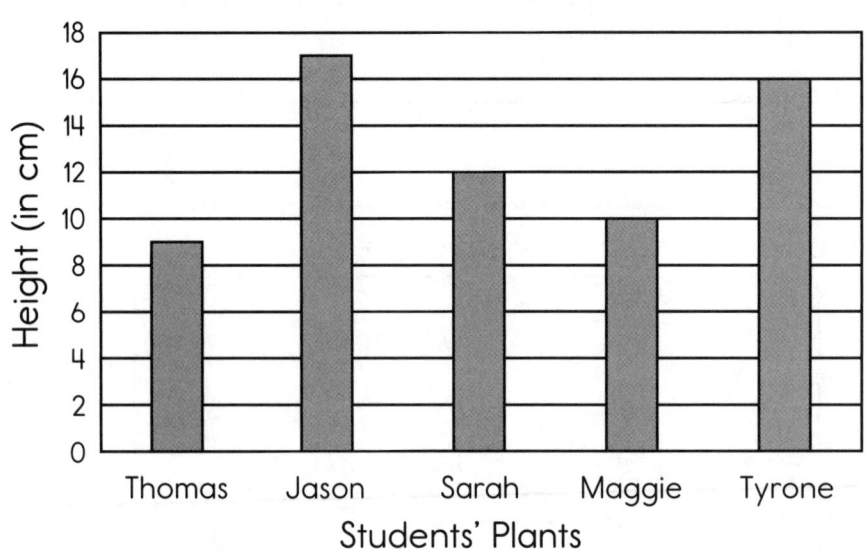

Heights of Students' Plants in Ms. Chung's Class

1. What are the students observing? _____

2. Who has the tallest plant? _____

3. Who has the shortest plant? _____

4. If Maggie's plant grows 2 cm, how tall will it be? _____ cm

5. If Sarah's plant grows 3 cm, will it be taller than Tyrone's? _____

6. Mr. Murphy's class has also been growing plants. They began their experiment one week after Ms. Chung's class did. Do you think his students' plants are taller or shorter than the plants in Ms. Chung's class? _____

7. Why? _____

_____

Logical Reasoning

Read the sentences in the box. Then, read each sentence below. Circle whether the sentence is **true** or **false**.

> All flowers are plants. Sunflowers are flowers.

1. Sunflowers are plants.   true   false

2. All plants are sunflowers.   true   false

Read the sentence in the box. Then, read each sentence below. Circle whether the sentence is **true** or **false**.

> Insects are a kind of bug.

1. All bugs are insects.   true   false

2. Insects are bugs.   true   false

The students in Ms. Gladd's class are all between 111 centimeters (cm) and 120 cm tall.

1. Darren is in Ms. Gladd's class. Which height could he be?

    A. 100 cm

    B. 115 cm

    C. 105 cm

    D. 132 cm

2. Which student could be in Ms. Gladd's class?

    A. Sofia (97 cm)

    B. Nathan (108 cm)

    C. Parker (121 cm)

    D. Lily (119 cm)

> Animals in the cat family have four legs, fur, and whiskers. They are meat eaters.

Write **C** next to each animal that is a cat. Write **N** next to each animal that is not a cat. Then, tell why it is not a cat.

1. _____ tiger   _____

2. _____ eagle   _____

3. _____ turtle   _____

Logical Reasoning

Answer the questions.

1 + 3 = 4	3 + 1 = 4
5 + 2 = 7	2 + 5 = 7
103 + 2 = 105	2 + 103 = 105

Look at the problems above. Use the same rule to answer the questions.

1. If 4 + 2 = 6, then 2 + 4 = _____

2. If 7 + 8 = 15, then 8 + 7 = _____

3. If 107 + 4 = 111, then _____ + 107 = 111

4. If 146 + 124 = 270, then 124 + 146 = _____

5. If 0 + 247 = 247, then 247 + _____ = 247

Look at the Venn diagram. Then, circle the best answer.

African Elephant   Asian Elephant

4,000-7,000 kg
large ears
male and females have tusks
2 fingers on tip of trunk
eat mostly leaves

has a trunk
4 legs
plant eater
mammal

3,000-6,000 kg
smaller ears
some males have tusks,
females do not
1 finger on tip of trunk
eat mostly grass

6. It has a trunk.   African elephant   Asian elephant   both

7. The female has tusks.   African elephant   Asian elephant   both

8. It eats mostly grass.   African elephant   Asian elephant   both

9. The local zoo just got a new female elephant. How might you know whether it is an African or Asian elephant?

_____

_____

Name_____

Properties of Matter

Matter is anything that takes up space and has mass.

1. Circle the objects made of matter.

   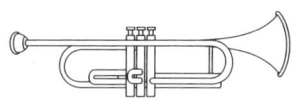 

Matter comes in all different sizes and masses.

1. Sort the following objects from smallest to largest:
   mouse, polar bear, human, and cat.

   _____,_____,_____,_____

2. Sort the following objects from lightest to heaviest:
   brick, candy bar, feather, and car.

   _____,_____,_____,_____

Matter can have different textures.

1. Name an object that is bumpy. _____

2. Name an object that is smooth. _____

3. Name an object that is squishy. _____

4. Name an object that is rough. _____

Matter has shape. Write the letter to match the object with its shape.

1. ____ 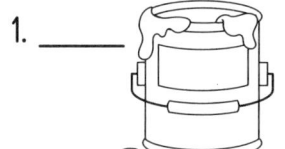   2. ____ 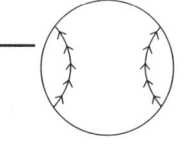   3. ____    4. ____

A. rectangular prism   B. cylinder   C. sphere   D. cone

Properties of Matter

Answer the questions.

1. Sort the objects in the word bank into two groups.

ice	soup	milk	fire	snow

2. Tell how you sorted the objects.

    _____

    _____

3. Pick two different objects. Using the properties of matter, write how you would tell a friend how they are alike and different.

    _____

    _____

    _____

    _____

    _____

Write the opposite of each property of matter.

4. long _____

5. hot _____

6. heavy _____

7. bumpy _____

1.RI.4, 1.SL.4, 1.L.1, 1.L.2, 1.L.5, 1.MD.C.4, 1.G.A.1

States of Matter

Fill in the blank with one of the states of matter: **solid**, **liquid**, or **gas**.

1. Water is a _____. It takes the shape of its container.

2. A wooden block is a _____. It does not take the shape of its container.

3. Oxygen is a _____. It fills the space it is given.

Day 1

Circle the state of matter to describe each object.

1. the air inside a balloon   solid   liquid   gas
2. an apple   solid   liquid   gas
3. rain   solid   liquid   gas
4. ice   solid   liquid   gas
5. water   solid   liquid   gas

Day 2

Complete the sentences. Unscramble the letters to fill in the missing words.

1. When ice melts, it changes to _____ (tewar).

2. When water freezes, it changes to _____ (cie).

3. When water boils, some of the water rises as water vapor, or

_____ (eamst).

Day 3

1. Tell one way to make an ice cube melt.

_____

_____

_____

_____

_____

Day 4

States of Matter

Answer the question.

1. Your class is talking about states of matter: solid, liquid, and gas. To help everyone understand, draw a picture showing each state of matter. Then, label each state of matter.

_____

_____

_____

1.RI.4, 1.SL.5, 1.L.4, 1.L.5

Sink or Float

Draw a line to match each word with its definition.

1. float

2. matter

3. sink

A. when an object stays on top of a liquid

B. when an object goes below the surface of a liquid

C. anything that takes up space and has mass

1. Name an object that would sink in water. Tell why it would sink.

_____

_____

2. Name an object that would float in water. Tell why it would float.

_____

_____

1. You want to conduct an experiment to test whether objects sink or float. What supplies do you need?

_____

_____

_____

1. Cara is doing an experiment to find whether objects sink or float. Read her steps below. Then, order the steps from **1–5**.

_____Last, I cleaned up.

_____Next, I collected my materials. I picked different objects to test.

_____Then, I recorded my predictions.

_____First, I cleaned off my work space.

_____I put each object in the water and recorded what happened.

Name_____

Sink or Float

Use the chart to answer the questions.

Object	Hypothesis: Do you think it will sink or float?	Results: Did it sink or float?
penny	float	sink
sponge	float	float
marble	sink	sink
toy boat	float	float
can of regular cola	sink	sink
can of diet cola	sink	float

1. How many objects did the student think would float? _____

2. How many objects did the student think would sink? _____

3. How many objects floated? _____

4. How many objects sank? _____

5. How many objects did the student predict correctly? _____

6. What is another object the student could test? _____

7. Do you predict it will sink or float? _____

8. Why? _____

_____

9. Draw a picture of a container of water. Add an object from the chart above that sank. Label it with an **S**. Add an object that floated. Label it with an **F**.

1.RI.1, 1.RI.3, 1.RI.4, 1.RF.4, 1.SL.5, 1.L.5, 1.L.6, 1.OA.C.5, 1.MD.C.4 CD-104812 • © Carson-Dellosa

Forms of Energy

Energy causes things to move.

1. List six things that move.

_____   _____   _____

_____   _____   _____

2. Name a way you could sort the items.

_____

Fill in the missing letters. Use the clues to figure out the three forms of energy.

1. L___gh___ energy comes from the sun and helps us to see objects.

2. Hea___ energy gives us warmth. It comes from sources like the sun and fire.

3. S___ ___ nd energy comes from vibrations. We use our ears to hear it.

For each picture, tell what each object needs to move or work.

1.    2.    3.    4.

_____   _____   _____   _____

1. Plants use energy from _____ to live and to grow.

A. animals   B. the sun   C. wind   D. other plants

2. Animals use energy from _____ to live and to grow.

A. the sun   B. animals   C. plants   D. animals and plants

Forms of Energy

Answer the questions.

Write the letter to match each object with the type of energy it uses.

1. _____ bear
2. _____ bus
3. _____ camp stove
4. _____ lamp
5. _____ kite
6. _____ tree

A.   electricity
B.   fire
C.   food
D.   gasoline
E.   sun
F.   wind

7. Name two energy sources a toy may need to move or work.

   _____   _____

8. Draw a picture showing two types of energy. Label the types of energy being used.

How Objects Move

1. Name the shape you think will roll down a hill the fastest.

_____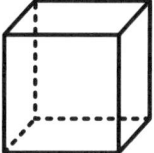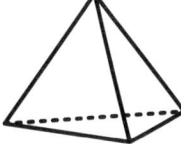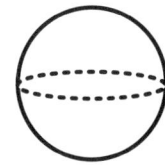

2. Why?

_____

Write the letter to match the type of motion with the picture.

1. _____ 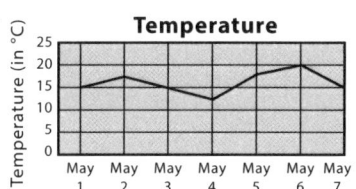 2. _____  3. _____

A. round and round in a circle   B. straight line   C. zigzag

1. Circle which animal can move faster.

   turtle   cheetah

2. Circle which object can move faster when pushed down a hill.

   sphere   cube

3. Circle which object can move faster when pulled.

   empty wagon   wagon full of bricks

1. Circle the ramp you think a toy car would move down the fastest.

   A. 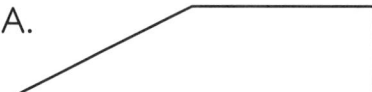   B.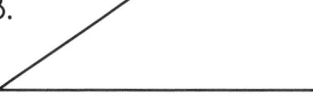

   C.

How Objects Move

Answer the questions.

1. You are talking to a friend about how things move differently. You want to help your friend understand. Draw a picture of a roller coaster that goes up and down and has a loop. Put an **F** near parts of the roller coaster where you think the car would go faster. Put an **S** near parts where you think the car would slow down.

Name_____

Force and Motion

1. If you drop an apple from your hand, it will fall to the ground because of _____.

   A. electricity

   B. gravity

   C. heat

   D. matter

1. Look at how the objects are being moved. Write **push** or **pull** under each picture.

   A.    B.    C.    D.

   _____   _____   _____   _____

1. Draw a picture of someone pulling an object. Draw an arrow showing which way the object will travel.

2. Draw a picture of someone pushing an object. Draw an arrow showing which way the object will travel.

1. If you were to push the ball, what would happen?

   _____
   _____

2. Why? _____

   _____
   _____

Force and Motion

Answer the questions.

1. Give an example of something you can push. _____

2. Give an example of something you can pull. _____

Tia and Brad did an experiment to test how high a ball bounces on different surfaces. Here is what they did.

Procedure

1. Hold a ball 1 meter from the testing surface.

2. Drop the ball.

3. Record how high the ball bounces back up.

Results

The ball bounced 50 cm on the classroom floor. It only bounced 30 cm on the carpet. The ball did not bounce at all on the pillow.

Surface	Height of Bounce (in cm)
classroom floor	
carpet	
	0

3. Fill in the missing words and numbers in the chart.

4. What surface did the ball bounce highest on? _____

5. What surface did the ball not bounce on? _____

6. How much higher did the ball bounce on the classroom floor than on the carpet? _____ cm

 1.RI.4, 1.RI.7, 1.SL.5, 1.L.5, 1.NBT.C.6, 1.MD.C.4 CD-104812 • © Carson-Dellosa

Name_____

Gravity

1. If you let go of a pencil you are holding, what do you think will happen to it?

_____

_____

_____

_____

Day 1

Fill in the blanks with words from the word bank.

down	Earth	gravity	moon

1. _____ is a force that pulls objects

toward the center of _____.

2. Because of gravity, what goes up must come _____.

3. The _____ revolves around Earth because of gravity.

Day 2

1. What are some questions you have about gravity?

_____

_____

2. What are some things that do not seem to be affected by gravity?

_____

_____

Day 3

Try to balance your pencil across your pointer finger.

1. Draw a picture to show how it balances best.

2. Why do you think it balanced best like this? Share your answer with a friend.

_____

Day 4

Gravity

Answer the question.

1. Imagine what Earth would be like if there was no gravity. Write a short story telling what school would be like without any gravity. Then, draw a picture of what your classroom might look like with no gravity.

_____

_____

_____

_____

_____

_____

_____

_____

1.RI.4, 1.W.3, 1.SL.1, 1.SL.5, 1.SL.6, 1.L.1, 1.L.2, 1.L.4 CD-104812 • © Carson-Dellosa

Name_____

Magnetism

1. Look around your classroom. List items you think are magnetic and items you think are not magnetic.

2. What is the same about the items in the Magnetic column?

_____

_____

Magnetic	Not Magnetic

Match each word with its definition.

1. attract

2. compass

3. magnet

4. repel

A. to push an object away from something

B. a magnetized tool that points north to help find direction

C. an object that attracts materials made of iron or steel

D. to pull an object towards something

1. What will happen if you hold a strong magnet next to a metal paper clip?

A. The paper clip will be pulled towards the magnet.

B. The paper clip will move away from the magnet.

C. Nothing will happen.

Circle whether the following facts are **true** or **false**.

1. If you set two magnets near each other, they may pull towards one another.   true   false

2. If you set two magnets near each other, they may push away from one another.   true   false

3. All metals are magnetic.   true   false

Magnetism

Answer the questions.

1. Circle the items you think are magnetic.

   metal door knob   soda can   nail clippers   crayon

   blade of a scissors   bottom of your shoe   nickel coin   cookie sheet

2. Why do you think they are magnetic?

   _____

Amy and John conducted an experiment to see which of their magnets was the strongest. Use their chart to answer the questions.

Magnet Shape	Number of Paper Clips It Held
bar	5
horseshoe	7
ring	3
rectangle	6
sphere	2

3. Use the chart to complete the bar graph.

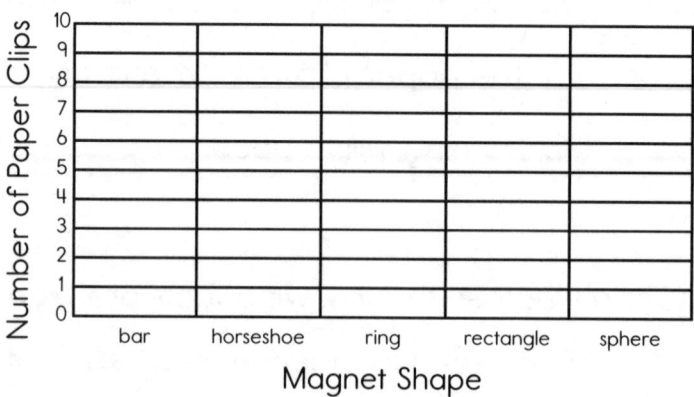

4. Which magnet held the most paper clips? _____

5. Which held the fewest? _____

6. How many more paper clips did the horseshoe magnet hold than the ring?

   _____

1.RI.4, 1.L.4, 1.L.5, 1.OA.A.1, 1.MD.C.4

Living and Nonliving Things

Write **P** for plant and **A** for animal.

1. _____ tree   2. _____ dog

3. _____ ladybug   4. _____ leaf

5. _____ banana   6. _____ person

Write **L** if the object is living. Write **N** if the object is nonliving.

1. _____ rock   2. _____ apple tree

3. _____ bird   4. _____ fish

5. _____ bike   6. _____ doll

1. What makes something living? Discuss your answer with a friend.

_____

_____

_____

_____

1. Write each word in the chart under the correct heading.

dragon	rock	snake	tree	unicorn	water

Living	Nonliving	Pretend

Living and Nonliving Things

Answer the questions.

Fill in the blanks with words from the word bank.

cars	elephants	plants	water

1. _____ does not move on its own but because of forces like gravity or wind. It is nonliving. Living things often live in it.

2. Even though _____ move, they are not living. They run on gasoline and not by themselves.

3. _____ are living because they grow and change. They get energy from the sun.

4. _____ are living because they breathe and grow. They get energy from food and are able to move on their own.

5. Circle the things that are living.

  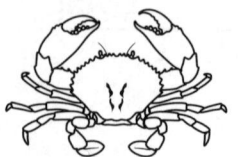

1.SL.6, 1.L.4, 1.L.5

Name_____

Plants

1. Circle the plants.

   apple tree   turtle   grass

   sunflower   lettuce   caterpillar

2. Give three more examples of plants.

   _____   _____   _____

The food we eat comes from different parts of a plant. Match the food with the part of the plant it is.

1. celery   A.   fruit

2. carrot   B.   leaf

3. grape   C.   root

4. lettuce   D.   stem

1. Circle the foods that have seeds.

   celery   apple   strawberries

   carrots   pumpkin   lettuce

2. Give three more examples of foods that have seeds.

   _____   _____   _____

Write **A** if the fact tells how sunflowers and apple trees are alike. Write **D** if the fact tells how they are different.

1. _____ has leaves   2. _____ has roots

3. _____ has bark   4. _____ makes seeds

5. _____ has branches

Plants

Answer the questions.

1. Label the parts of the flower with words from the word bank.

leaf	petal	root	stem

A. _____

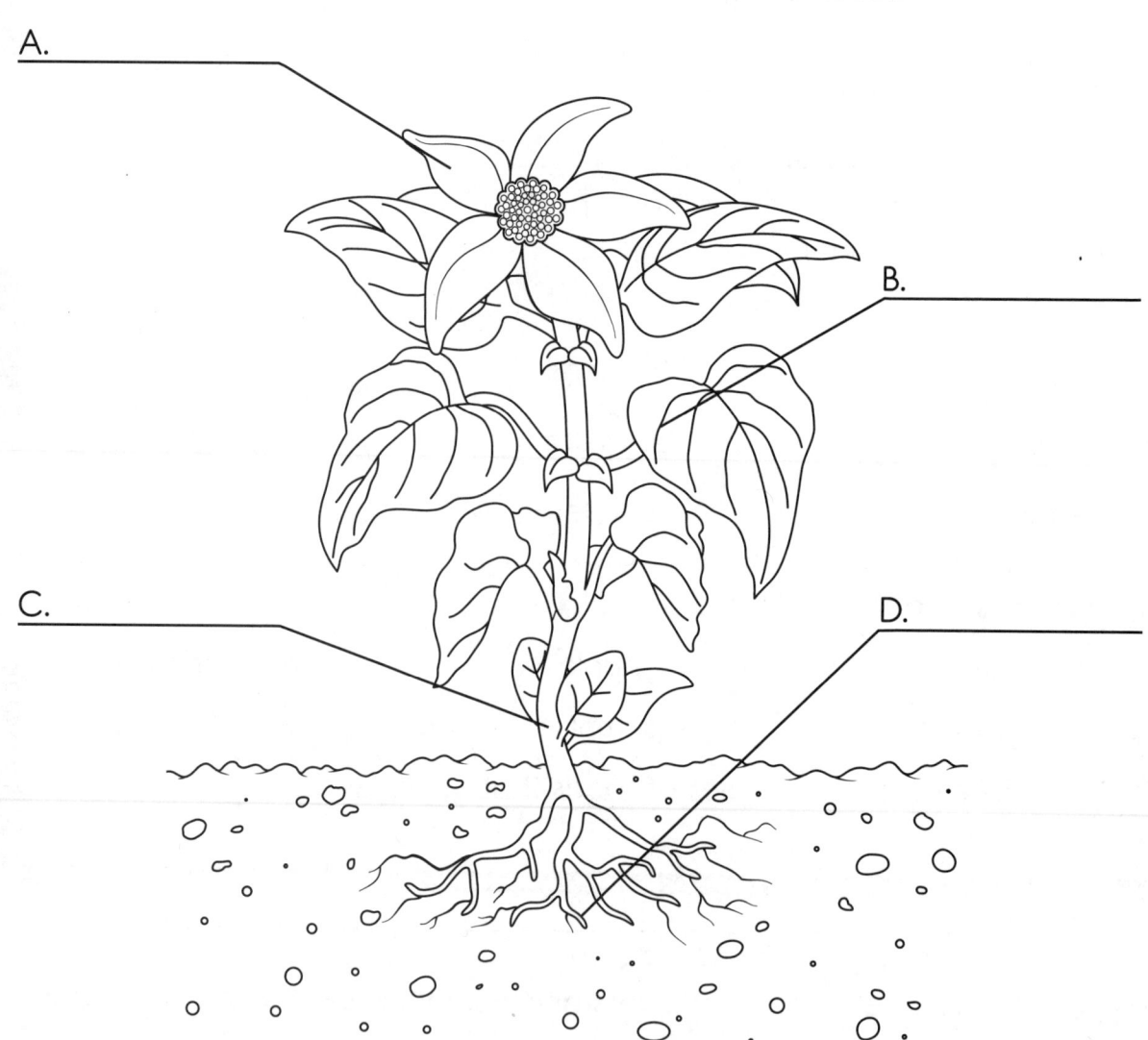

B. _____

C. _____

D. _____

2. What is your favorite vegetable? _____

3. What part of the plant does it come from? _____

Animals

Day 1

1. Most mammals have 4 legs. Spiders have 8 legs. So, 2 mammals and 1 spider would have _____ legs altogether.

   A. 12

   B. 14

   C. 16

   D. 18

Day 2

1. Circle the animals that hatch from eggs.

   birds   turtles   people   dogs

2. Circle the animals that fly.

   bats   whales   birds   people

Day 3

Use the words from the word bank to complete the sentences.

| dog | penguin | frog | shark | turtle |

1. A _____ is a mammal.

2. A _____ is an amphibian.

3. A _____ is a reptile.

4. A _____ is a bird.

5. A _____ is a fish.

Day 4

Use a verb, or action word, to tell how each animal moves.

1. snake _____

2. bird _____

3. kangaroo _____

4. cheetah _____

Name_____

Animals

Answer the questions.

Name and draw a kind of shelter for each animal. Discuss your drawings with a friend.

1. bear _____

2. bird _____

3. bee _____

4. human _____

5. mole _____

6. beaver _____

1.RI.1, 1.RI.7, 1.SL.5, 1.L.4, 1.L.5, 1.OA.A.1, 1.OA.A.2

Plant and Animal Needs and Interdependence

1. Name one way the sun is helpful to plants or animals.

   _____

2. Name one way the sun is harmful to plants or animals.

   _____

Circle whether each sentence is **true** or **false**.

1. Animals help plants through pollination.   true   false

2. Plants help animals by making oxygen.   true   false

3. Animals give off carbon dioxide that plants
   use to make food.   true   false

Complete the sentences. Unscramble the letters to fill in the missing words.

1. Plants may be used for shelters. Birds can make _____
   (esnts) with plants.

2. Insects sometimes use plants to blend in. Green grasshoppers can
   _____ (idhe) in grass.

3. Many animals use plants as a food source. Pandas _____
   (eta) bamboo.

Fill in the blanks with words from the word bank.

insects	leaves	plants	soil

1. Worms break down dead leaves into _____.

2. Seeds that are replanted become new _____.

3. Animals plant parts like fruits and _____ .

4. _____ help to pollinate flowers.

Plant and Animal Needs and Interdependence

Answer the questions. Then, share your answers with a partner.

1. Explain why plants need sunlight.

   _____

   _____

2. Explain why animals need plants.

   _____

   _____

3. Explain why plants need animals.

   _____

   _____

At breakfast, Maddie observed the animals in her yard. She saw 3 squirrels. Each grabbed 1 acorn to eat. A bird also ate 1 acorn. Then, 2 more squirrels each ate 1 acorn.

4. How many acorns did Maddie see animals eat? _____

5. Later that morning, she saw 3 more squirrels eat 1 acorn each. How many acorns did she see eaten now? _____

6. By the end of the day, Maddie had seen 15 acorns eaten by animals. How many acorns were eaten in the afternoon? _____

1.RI.3, 1.RI.4, 1.SL.4, 1.L.4, 1.L.5, 1.OA.A.1, 1.OA.A.2, 1.OA.B.4

Plant and Animal Structures

Draw a line to match the animal with the body part it has that helps it survive.

1. duck

2. fish

3. grasshopper

4. whale

A. strong back legs

B. webbed feet

C. layer of blubber

D. gills

1. Where might a green insect be able to hide?

_____

2. Why would this be a good place?

_____

3. Why might it need to hide?

_____

Fill in the blanks using words from the word bank.

bee	birds	flowers	skunks

1. A _____ has a stinger to protect it from predators.

2. _____ have colorful petals to attract insects.

3. Some _____ have strong beaks to crack open seeds.

4. _____ are black and white to warn other animals.

Circle the correct answer.

1. What do animals with sharp teeth often eat?   meat   plants

2. Which animal has sharper teeth?   horse   lion

3. What do animals with flat teeth often eat?   meat   plants

4. Which animal has flatter teeth?   cow   tiger

Plant and Animal Structures

Answer the questions.

1. Cactus plants have a thick, waxy skin that stores water inside. This is useful in climates with little rainfall. Where would cactus plants most likely be found?

   A. desert

   B. ocean

   C. rain forest

2. Your class is discussing how animals survive. Draw an animal to explain your thoughts. Label at least two features its body has that help it survive.

Fill in the blanks with words from the word bank.

beak	hibernate	hide	roots

3. A snowshoe hare's white fur helps it to camouflage, or
   _____, in the snow.

4. A hummingbird's long _____ helps it collect nectar from flowers.

5. Some animals _____ to survive without food in the winter.

6. Trees adapt to long periods without water by growing long
   _____.

Traits of Living Things

Write the letter to match the baby with the adult.

1. ___    2. ___    3. ___

A.    B.    C.

Day 1

1. Some children have the same hair color or eye color as one or both of their parents. Others share the trait of freckles. What are some features you share with your parents?

_____

_____

_____

Day 2

1. What are some things that might be different about a child and his parents?

_____

_____

_____

_____

Day 3

1. Emma wonders if plants of the same kind are exactly the same. She planted three seeds from the same sunflower plant. Circle what you think she found to be the same as the parent plant.

size of the new plants   shape of the leaves

number of leaves   the seeds all grew into sunflower plants

Day 4

Traits of Living Things

Answer the questions.

1. What are some ways humans are different from one another?

    _____

    _____

2. What are some ways humans are all alike?

    _____

    _____

Kristen's dog, Muffy, just had puppies. Kristen made a chart showing what the puppies look like. Use the chart to answer the questions.

Puppy's Name	Gender	Fur Color	Coat Length	Tail
Muffin	female	black	long	straight
Pancake	male	brown	short	straight
Luna	female	black	short	straight
Jack	male	brown	short	straight
Rosie	female	brown	long	straight

3. How many puppies did Muffy have? _____

4. How many of the puppies are male? _____

5. Muffy has short brown fur. How many of the puppies have the same fur color as Muffy? _____

6. How many puppies have a different coat length than Muffy has? _____

7. What do all of the puppies have in common? _____

8. What might the father of the puppies look like?

    _____

9. Why do you think that?

    _____

1.RI.3, 1.SL.4, 1.SL.6, 1.OA.A.1, 1.MD.C.4

Habitats

1. Name some features that would help an animal survive in a cold environment.

   _____

   _____

   _____

   _____

Name the habitat that is being described: **desert**, **tundra**, or **rain forest**.

	Animals	Plants	Climate	Habitat
1.	arctic hares, polar bears	arctic moss	very cold, dry	
2.	lizards, coyotes	prickly pear cactus	often hot, little rain	
	monkeys, tree frogs	rubber tree	very rainy	

Match the description with the body of water.

1. moving freshwater   A.   lake

2. large, freshwater, surrounded by land   B.   pond

3. smaller than a lake, freshwater,   C.   river
   surrounded by land

Jack has been observing the prairie by his school. Here is what he saw:

   Plants: pasture rose, milkweed, prairie violet, coneflower
   Animals: squirrel, sparrow, cricket, hawk, vole, turtle, beetle, butterfly

1. How many different kinds of plants did Jack see? _____

2. How many different kinds of animals did he see? _____

3. How many total kinds of plants and animals did he see? _____

Habitats

Answer the questions.

Match the creature to its habitat.

1.    A.

2.    B.

3.    C.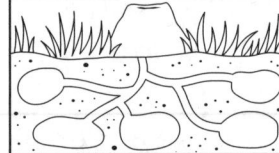

4. 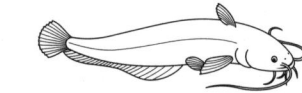   D.

5.    E.

6. Choose one animal and tell why its home meets the animal's needs. Share your thoughts with a partner.

_____

_____

_____

_____

_____

Name_____

Life Cycles

1. Number the pictures in the order that an egg would become a butterfly.

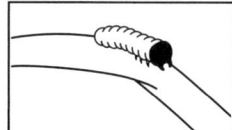 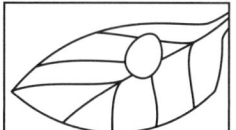   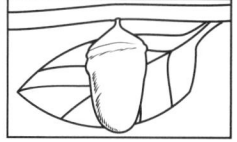

_____   _____   _____   _____   _____

1. Write the numbers **1** to **5** to put the life cycle of a frog in order.

_____ A frog lays eggs in the water.

_____ Their front legs grow, and their tails get shorter. The gills are now covered. Their lungs being used. They are now adult frogs.

_____ A female adult frog lays eggs. The cycle starts over.

_____ Next, the tadpoles start to grow back legs. Their lungs grow.

_____ Then, the eggs hatch. The tadpoles each have a tail and gills.

1. Use the word bank to label the pictures of the life cycle of a pumpkin.

| flower | orange pumpkin | seed | vine |

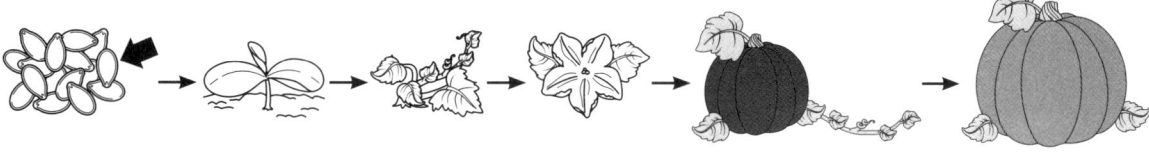

_____   sprout   _____   _____   green pumpkin   _____

1. What are some changes a human goes through when growing from a baby to an adult?

_____

_____

_____

_____

Life Cycles

Use the diagram to answer the questions.

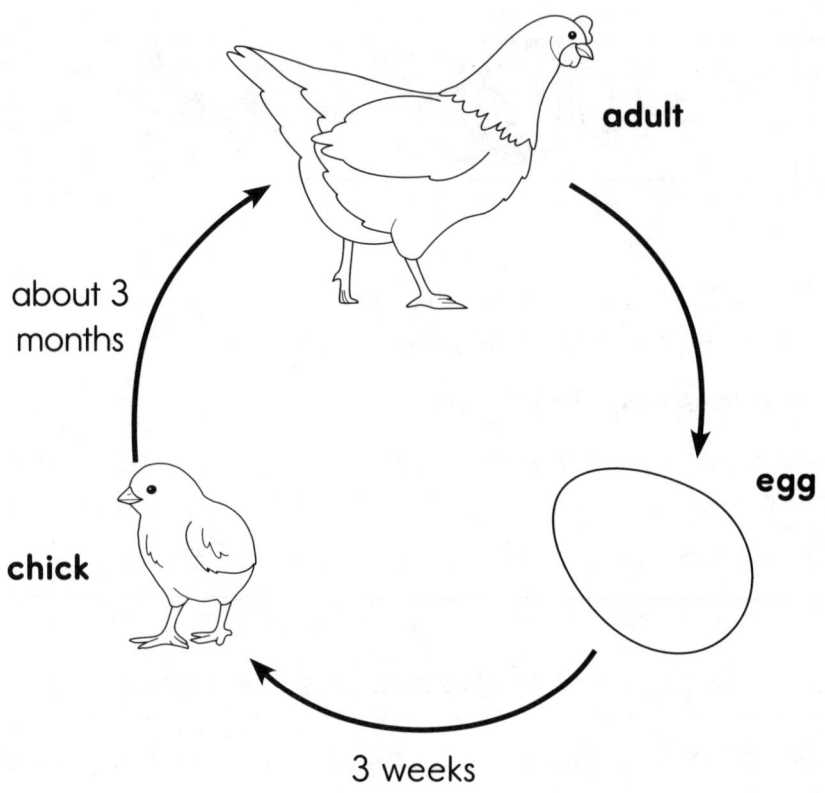

about 3 months

adult

egg

chick

3 weeks

1. What is a baby chicken called? _____

2. How long does it take for the egg to hatch? _____

3. What does the arrow that is pointing from the adult to the egg mean?

   _____

4. An egg was laid 2 weeks ago. About how much longer until it hatches?

   _____

The Human Body

Day 1

1. The bark of the tree is similar to what part of a person? _____

2. Why? _____
_____

3. Tell one other part of your body that is similar to a tree and why.

_____

Day 2

1. What are some things on your body that grow or change?

_____

_____

_____

_____

Day 3

You have 2 bones in each thumb and 3 bones in each of your other fingers.

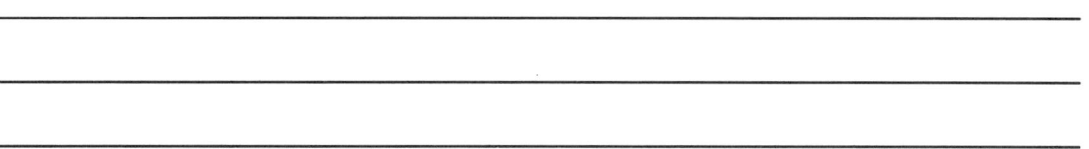

1. How many bones are in one hand? _____

2. How many bones are in both hands? _____

3. Explain how you found the answer. _____

Day 4

Fill in the blanks with words from the word bank.

bones	heart	lungs	muscles

1. Your _____ give your body its structure.

2. You use your _____ to move body parts.

3. Your _____ are used to breathe.

4. Your _____ pumps blood through your body.

The Human Body

Use the chart to answer the questions. Show each answer as a number sentence using <, >, or =.

1. Do humans have more or less permanent teeth than cats?

   _____ 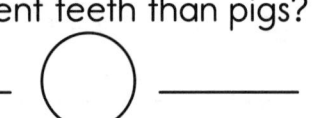 _____

2. Do humans have more or less permanent teeth than pigs?

   _____ ◯ _____

Animal	Number of Permanent Teeth
human	32
cat	30
dog	42
pig	44

3. What animal in the chart has the most permanent teeth?

   _____

4. Label the body parts on the person.

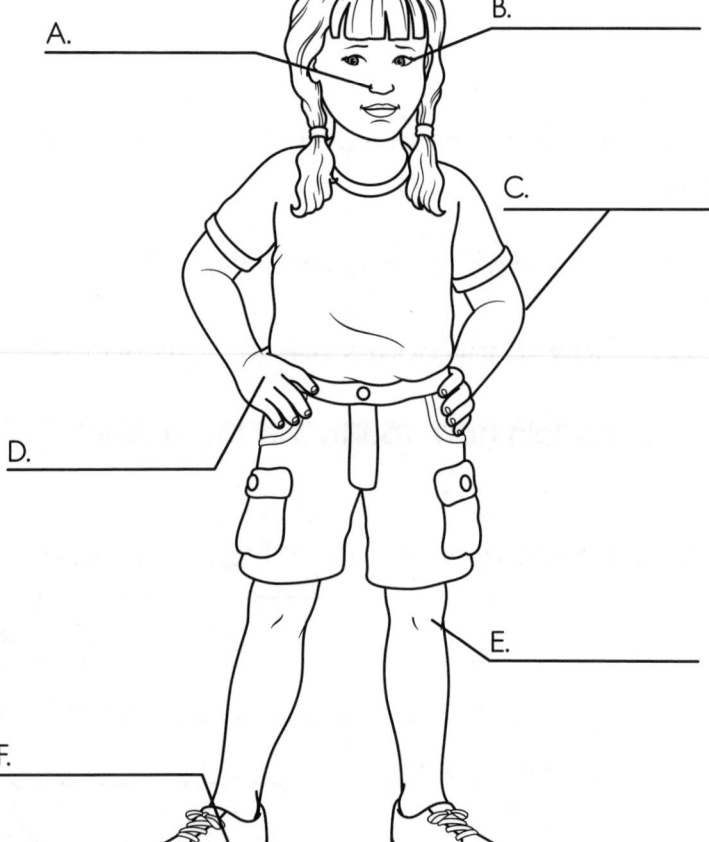

A. ____

B. ____

C. ____

D. ____

E. ____

F. ____

Day and Night

1. If it is daytime on one side of the world, it is _____ on the other side of the world.

2. Will you be more likely to be in school at noon or at midnight? _____

3. What is an animal that is awake at night? _____

4. What will that animal probably do during the day? _____

Next to each time, write whether it is day or night.

1. 8:30 pm _____   2. 10:00 am _____

3. noon _____   4. midnight _____

5. 3:30 pm _____

Write **T** for true or **F** for false.

1. _____ The sun circles around Earth to make day and night.

2. _____ Earth circles around the sun to make day and night.

3. _____ Earth rotates around on its axis to make day and night.

4. _____ When it is day on one side of Earth, it is night on the other side of Earth.

1. Write the animals into the chart. Some words may be used more than once.

| bat | butterfly | cardinal | moon | owl | stars | sun |

Things I Can See During the Day	Things I Can See at Night

Day and Night

Answer the questions.

1. If it is 6:00 pm on one side of the world, what time is it on the exact other side of the world?

   A. 12:00 pm   B. 12:00 am   C. 6:00 am   D. 6:00 pm

2. If it is 11:00 pm, what time is it a half hour later?

   A. 12:00 pm   B. 11:30 pm   C. 11:30 am   D. 11:00 am

3. Your class is talking about day and night. Draw a picture to show what you are most likely doing at noon. Label your picture.

4. Your class is talking about day and night. Draw a picture to show what you are most likely doing at midnight. Label your picture.

1.RI.3, 1.RI.4, 1.SL.5, 1.L.4, 1.L.5, 1.MD.B.3   CD-104812 • © Carson-Dellosa

Name_____

The Seasons

1. Name and draw two items people wear on a cold winter day.

   _____ _____

2. Name and draw two items people wear on a hot summer day.

   _____ _____

Day 1

Name what season each month falls in.

1. January _____

2. July _____

3. October _____

4. April _____

Day 2

In complete sentences, tell an activity you do in each season.

1. spring _____

2. summer _____

3. autumn _____

4. winter _____

Day 3

1. What season comes before winter? _____

2. What season comes after spring? _____

3. What season comes before autumn? _____

4. What season comes after summer? _____

Day 4

The Seasons

Answer the questions.

Write the name of the season under each picture.

1. _____

2. _____

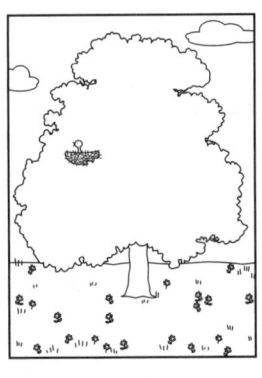

3. _____

4. _____

In complete sentences, tell something that happens in nature during each season. Share your answers with a friend.

5. spring _____

_____

6. summer _____

_____

7. autumn _____

_____

8. winter _____

_____

1.RI.3, 1.RI.4, 1.SL.6, 1.L.2, 1.L.5

Earth and the Sun

Fill in the blanks with words from the word bank.

| rays | skin | sun | sunblock |

1. When you go out in the _____, it is important to protect your _____.

2. The sun's _____ can cause sunburns.

3. Wear _____ to prevent your skin from burning.

It takes Earth one year to orbit the sun.

1. How many months is one orbit? _____

2. How many days is one orbit? _____

3. How is Earth's orbit a cycle?

_____

Fill in the missing vowels to complete facts about the sun.

1. The sun provides l__ght to help us see.

2. It provides h__ __t to keep animals warm during the day.

3. Plants use energy from the sun, along with carbon dioxide and water, to make f__ __d.

4. Solar energy can be turned into electricity. We can use that power for electronics, such as c__mput__rs.

Circle if the following sentences are **true** or **false**.

1. The sun is cold.   true   false

2. The sun is a star.   true   false

3. The sun is smaller than Earth.   true   false

4. The sun is closer to Earth than the moon.   true   false

5. The sun is important to all living things on Earth.   true   false

Earth and the Sun

Answer the questions.

The children in the picture are doing an experiment with the sun.

1. What question do you think they are trying to answer?

   _____

2. What do you think they will find out?

   _____

3. What other equipment might they need?

   _____

Circle which of the three suns is correct in each picture.

4.

5.

6. How do you know? Use the picture to explain your answers to a friend.

   _____

Name_____

The Moon

1. How many moons does Earth have?

   _____

2. Use three words to describe what the moon looks like.

   _____ _____ _____

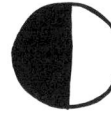

1. What picture should the missing moon look like?

    A.    B.    C.

1. Put the following in order from smallest to largest: Earth, moon, sun.

   _____ , _____ , _____

2. How do you know?

   _____

1. Why can we see the moon in the night sky?

   _____

   _____

   _____

   _____

   _____

The Moon

Answer the questions.

Match the picture with the shape the moon appears to be.

1.
   A. circle

2.
   B. crescent (like a fingernail)

3.
   C. half circle

4. The moon is actually what 3-D shape?

   A. circle   B. cube

   C. pyramid   D. sphere

Circle whether each fact is true of the **moon**, the **sun**, or **both**.

5.	sometimes can be seen at night and sometimes during the day	sun	moon	both
6.	is smaller than Earth	sun	moon	both
7.	is a star	sun	moon	both
8.	can be seen at different places in the sky at different times	sun	moon	both
9.	travels around Earth	sun	moon	both
10.	makes its own light	sun	moon	both
11.	is closer to Earth	sun	moon	both

1.RI.4, 1.RI.7, 1.L.4, 1.MD.A.1, 1.MD.C.4, 1.G.A.1, 1.G.A.2

Weather

1. Which temperature is most likely on a hot day?

   A. 0°C (32°F)

   B. 10°C (50°F)

   C. 27°C (81°F)

   D. 80°C (176°F)

2. Which temperature is most likely on a cold day?

   A. 0°C (32°F)

   B. 10°C (50°F)

   C. 27°C (81°F)

   D. 80°C (176°F)

1. Describe the weather this morning.

   _____

   _____

   _____

   _____

Use the graph to answer the questions.

1. What day had the most rainfall?

   _____

2. What days had no rainfall?

   _____

3. What is the total rainfall for the weekend? _____ cm

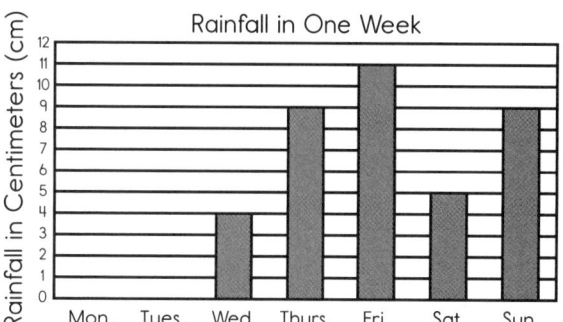

Rainfall in One Week

In the morning, the temperature was 14°C. By afternoon, the temperature was 20°C.

1. How many degrees did the temperature change? _____ °C

2. How did you get your answer?

   _____

Weather

Answer the questions.

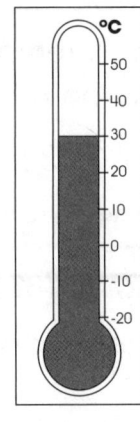

1. What is the temperature in degrees Celsius?
   _____°C

Write a word to describe the weather in each picture.

2. _____

3. _____

4. _____

5. _____

Earth's Surface

1. Circle which of the following can be found on Earth's surface.

   water   rocks   minerals   soil

2. Give three more examples of things that can be found on Earth's surface.

   _____ _____ _____

Circle if the following sentences are **true** or **false**.

1. Water shapes the land.   true   false

2. Household products are sometimes made from rocks and minerals.   true   false

3. There is only one kind of soil.   true   false

Earth's surface is shaped by different forces. Complete the words to learn three ways it is shaped.

1. W__t__r changes the shape of the land. Ocean waves crash on the shore. Rivers cut away at the land and smooth rocks.

2. A glacier is a large mass of i___e that moves slowly. It can move rocks as it travels and scrape the ground beneath it.

3. W__nd can be a powerful force even though you cannot see it.

1. Sort the following things found on Earth into the chart.
   water   trees   rocks   birds   minerals   grass   soil

Living	Nonliving

Earth's Surface

Answer the questions.

Use two words to describe each of the items found on Earth's surface.

1. sand _____ _____

2. mountain _____ _____

3. water _____ _____

4. soil _____ _____

5. rock _____ _____

6. grass _____ _____

Fill in the blanks with the words from the word bank.

beavers	roots	soil	tunnels	water

7. Plant _____ can break through rocks.

8. _____ can seep between rocks. When it freezes, it expands. It can crack the rocks apart.

9. Earthworms create new _____ by eating dead plants.

10. People build _____ through the mountains for trains and roads.

11. _____ build dams. The dams change waterways.

1.RF.3, 1.L.1, 1.L.4, 1.L.5

Water

1. Name three different bodies of water.

    _____   _____   _____

2. Tell about a body of water you have visited.

    _____

    _____

Fill in the missing letters to complete each sentence.

1. Water that falls from clouds as a liquid is called r___ ___n.

2. Water that falls from clouds as a crystal is called sn___ ___.

3. Tiny drops of w___t ___r form clouds.

4. Falling water collects in str___ ___ms and other bodies of water.

Terra County Landforms			
ponds	2	valleys	1
streams	4	lakes	1

1. Circle the sources of water.

2. How many water sources were found in all? _____

Circle the correct word to complete each sentence.

1. Earth's surface is mostly made up of _____(land, water).

2. Most of Earth's water is _____(salty, freshwater).

3. Some of Earth's water cannot be used because it is
    _____(frozen, fresh).

4. Water found beneath Earth's surface is called
    _____(groundwater, tap water).

Water

Answer the questions.

1. In a complete sentence, name three ways humans or animals use water.

   _____

   _____

2. Water becomes a solid when it _____.

   A. evaporates

   B. freezes

   C. melts

3. Ice becomes a liquid when it _____.

   A. evaporates

   B. freezes

   C. melts

4. Water that evaporates into the air is called _____.

   A. ice

   B. water

   C. water vapor

5. Put the following bodies of water in order from smallest to largest.

   | lake | ocean | pond | puddle |

   _____, _____, _____, _____

1.RI.4, 1.RF.3, 1.W.3, 1.W.8, 1.L.1, 1.L.2, 1.OA.A.1, 1.OA.A.2 CD-104812 • © Carson-Dellosa

Name_____

Information Technology

Match the definition to the name of the computer part.

1. moved with your hand to control what is on the computer screen

2. the screen of the computer

3. used to type words or numbers

4. worn on your ears to hear sounds on the computer

A. keyboard

B. headphones

C. monitor

D. mouse

Day 1

1. What is better about using a computer than writing something by hand?

_____

2. What is better about writing something by hand than using a computer?

_____

Day 2

1. How can a computer be used as a science tool? Share your answer with a friend.

_____
_____
_____
_____

Day 3

1. What are some rules to follow when using a computer?

_____
_____
_____
_____

Day 4

Information Technology

Answer the questions.

font	print	save	scroll

Fill in the blanks with the words from the word bank.

1. When you are done typing something you want to keep, be sure to _____ your work.

2. You can use the _____ bar to move up or down on a page.

3. If you would like to have a paper copy of something, you can _____ your work.

4. The size and style of letters and numbers on a computer is called the _____.

5. You are a scientist. You need a computer for your job. Write a letter to your boss explaining why a computer will help you do your job.

_____

_____

_____

_____

_____

_____

_____

_____

_____

_____

_____

1.RI.3, 1.RF.3, 1.W.2, 1.SL.4, 1.SL.6, 1.L.1, 1.L.4

Engineering Design

1. What are some tools you would need to build a wooden bird feeder?

   _____

2. What are some tools you would need to make a paper airplane?

   _____

3. Will you need the same tools for both? Why or why not?

   _____

Day 1

Your class has been building toy cars from empty containers.

1. How can you test whose car is fastest?

   _____

2. What equipment would you need?

   _____

Day 2

1. Pick a household object that could be improved. Draw and write about how you could make it better.

   _____

   _____

   _____

   _____

Day 3

What are some new uses for the following?

1. a paper clip _____

2. an empty tissue box _____

3. a toilet paper tube _____

Day 4

Engineering Design

Answer the questions.

1. Create two designs for a craft-stick bridge that could hold weights on top of it.

2. Which of your two designs do you think would be stronger? Circle it.

3. What are some shapes you used in your designs?

   _____

4. Tell why you think the design you circled would be stronger.

   _____

   _____

   _____

1.W.1, 1.SL.5, 1.SL.6, 1.L.1, 1.L.2, 1.G.A.2

Name_____

Healthy Habits

1. List three ways to be healthy.

    A. _____

    B. _____

    C. _____

1. Sort the foods Carrie ate into the different food groups.

| banana | carrot | cheese | eggs | lettuce | milk |
| rice | strawberries | walnuts | whole wheat roll | | |

Fruits	Vegetables	Grains	Protein	Dairy

1. Fill in the blanks with words from the word bank.

| bathing | brush | floss | throw away | wash |

There are many ways to be healthy. For healthy gums and teeth, it is important to _____ and _____ your teeth every day. _____ your hands with soap and warm water to stay healthy. _____ used tissues right away. _____ helps to keep your body clean.

1. Name two times you should wash your hands.

    _____

    _____

    _____

    _____

Healthy Habits

Answer the questions.

1. Explain different ways to stay healthy. Then, tell why being healthy is important.

   _____

   _____

   _____

   _____

2. What are some physical activities you enjoy doing?

   _____

3. What are some healthy foods you enjoy eating?

   _____

4. You want to talk to your friends about making healthy choices. Create a poster to help you convince them to have healthy habits.

1.RF.3, 1.W.2, 1.W.8, 1.SL.5, 1.L.1, 1.L.4, 1.L.5, 1.OA.A.1

Name_____

Resources

Day 1

1. Describe three way humans use plants in their everyday lives.

    _____

    _____

    _____

    _____

    _____

Day 2

1. Draw a circle around each item that comes mainly from plants.

2. Draw a triangle around each item that comes mainly from animals.

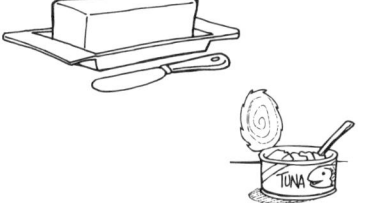  

Day 3

Write **R** in front of the things that are renewable resources. Write **N** in front of the things that are nonrenewable resources.

1. _____ solar energy (energy from the sun)   2. _____ gasoline

3. _____ coal   4. _____ wind energy

5. _____ trees

Day 4

1. Circle the resources that you used today.

    gasoline   coal   wind   solar energy   plants   water

2. Choose one resource you circled. Tell how you used it.

    _____

    _____

Resources

Answer the questions.

1. Write a paragraph telling about nonrenewable resources. Give examples of how we can use less of them.

    _____

    _____

    _____

    _____

    _____

2. Draw a picture to show two ways a tree is used by people or animals.

    ┌───┐
    │   │
    │   │
    │   │
    │   │
    │   │
    │   │
    │   │
    └───┘

3. Is a tree a renewable or nonrenewable resource? _____

4. Why?

    _____

1.RI.3, 1.RI.4, 1.W.2, 1.SL.4, 1.SL.5, 1.L.1, 1.L.5   CD-104812 • © Carson-Dellosa

Human Impact on Earth

Write **P** if the action affects plants. Write **A** if the action affects wild animals. Write **N** if the action affects natural resources. Each sentence may have more than one letter.

1. _____ Trees are planted.

2. _____ Crops are sprayed with a chemical to keep insects away.

3. _____ Land is cleared to build a new shopping center.

4. _____ Solar panels are installed on top of a roof to collect energy.

A low-flow toilet uses about 6 liters of water each time it is flushed. A toilet that is not low-flow may use about 13 liters each time it is flushed.

1. How many liters of water will each type of toilet use if flushed 2 times in a day?

low-flow: _____ liters   not low-flow: _____ liters

3. How many liters of water would be saved each day with the low-flow toilet? _____ liters

Look at the picture.

1. What is happening in the picture?

_____

_____

2. Who or what might be affected?

_____

Unscramble the letters to fill in the missing words.

1. Riding a bike instead of taking a car to work reduces _____ (iar) pollution.

2. Fertilizers used on lawns can trickle into the _____ (tewar) supply.

3. Be sure to throw garbage into trash cans instead of on the ground. Littering pollutes our _____ (ndal).

Day 1

Day 2

Day 3

Day 4

Human Impact on Earth

Answer the questions.

1. Tell ways humans help Earth and ways they harm it.

    _____

    _____

    _____

    _____

    _____

2. Design a bumper sticker for a car with a message about helping our
    planet.

1.RI.4, 1.RI.7, 1.RF.3, 1.W.2, 1.L.1, 1.L.4, 1.OA.A.1   CD-104812 • © Carson-Dellosa

Conservation

1. Give an example of something you can recycle.

   _____

2. Give an example of something you cannot recycle.

   _____

1. Tell one way humans have helped Earth through conservation.

   _____

   _____

   _____

   _____

| **Reuse:** to use again | **Reduce:** to use less of something |
| **Recycle:** to make something new from a used material | |

Circle the best answer.

1. using old paper as scrap paper   reducing   reusing   recycling
2. taking shorter showers   reducing   reusing   recycling
3. making toys from plastic bottles   reducing   reusing   recycling

1. Tell one way you are conserving resources at home.

   _____

2. Tell one way you are conserving resources at school.

   _____

3. Tell one way you are conserving resources outside.

   _____

Conservation

Answer the questions.

1. Write the following items under the correct heading.

    | banana peel | glass bottle | newspaper |
    | slice of pizza | soda can | foam takeout box |

Can Be Recycled	Cannot Be Recycled

2. Write a paragraph on ways your school could use fewer resources.

    _____

    _____

    _____

    _____

    _____

    _____

    _____

    _____

    _____

    _____

1.RI.3, 1.RI.4, 1.W.2, 1.SL.4, 1.L.1, 1.L.5

Page 9

Day 1: 1. 3; 2. 5; **Day 2:** 1. D; 2. C; 3. B; 4. A;
Day 3: 1. u, e (ruler); 2. r, c (microscope);
3. a, e (beaker); 4. o, e (goggles); **Day 4:** 1.
Answers will vary but may include a camera, net,
hand lens, beaker, binoculars, or notebook.

Page 10

1. C; 2. A. beaker; B. hand lens; C. binoculars;
3. stick B; 4. Answers will vary but may include to
track plant or animal growth or change
over time.

Page 11

Day 1: 1. Answers will vary but may include by
length, width, or edge type. **Day 2:** 1. Can Fly;
2. Cannot Fly; **Day 3:** 1. 17; 2. 16; **Day 4:** 1. Answers
will vary.

Page 12

1. Answers will vary. 2. Answers will vary but may
include sorting by number of legs, size, kind of
animal, or whether the animal has fur, feathers,
or scales.

Page 13

Day 1: 1. 18; **Day 2:** 1. 24 hours; 2. 30 seconds;
3. once a week; **Day 3:** 1. 1; 2. meter; **Day 4:** 1. the
counters; 2. Answers will vary but may include
that the side of the scale with the heavier object
will be lower than the other side.

Page 14

1. 10; 2. flea, bumblebee, grasshopper, luna moth;
3. A. Answers will vary but may include a ruler,
yardstick, meterstick, or tape measure. B. balance
scale; C. thermometer; D. clock, stopwatch, or
timer; 4. A. grams; B. seconds; C. centimeters;
D. Celsius.

Page 15

Day 1: 1. Answers will vary. **Day 2:** 1. Answers
will vary but may include making a prediction,
gathering materials, keeping track of results,
testing, and sharing results. **Day 3:** 1. The bridges
were tested with different types of coins and
they can't be compared. **Day 4:** 1. Does salt
water or freshwater freeze faster? 2. a container
of salt water, a container of freshwater, a
freezer, and a timer; 3. Answers will vary.

Page 16

1. black oil sunflower seeds; 2. safflower seeds;
3. no; 4. 7 seeds ; 5. 16 seeds

Page 17

Day 1: 1. B; **Day 2:** 1. Answers will vary but may
include to wash off chemicals, soil, or other
materials from your hands. **Day 3:** 1. B; 2. C;
3. A; 4. D; **Day 4:** 1. hazard, poison, flammable

Page 18

1. Answers will vary but might include that if the
spill was on the floor, someone could slip and fall.
2. hair; 3. food; 4. drink; 5–6. Answers will vary.

Page 19

Day 1: 1. smell; 2. see; 3. hear; 4. taste; 5. touch;
Day 2: 1–4. Answers will vary. **Day 3:** 1–4. Answers
will vary. **Day 4:** 1. Answers will vary but may
include using the sense of sight to see plants
and animals; using the sense of hearing to hear
animals; using the sense of touch to feel plants,
rocks, and the temperature of the air; and using
the sense of smell to detect the odors of plants
and animals.

Page 20

1. Answers will vary but should include all five
senses. 2. Answers will vary but should include all
five senses. 3. Answers will vary.

Page 21

Day 1: 1. telescope, pencil, camera, notebook;
Day 2: 1. h (who); 2. h, a (what); 3. e, e (where); 4. h, n (when); 5. o, w (how); 6. h (why); 7. i (which); **Day 3:** 1. Answers will vary.
Day 4: 1. Answers will vary but may include what time are insects outside? What do insects eat?

Page 22

1. Answers will vary. 2. Answers will vary but may include looking on the Internet or in books or by performing an experiment. 3. Answers will vary but may include creating a chart.

Page 23

Day 1: 1. O; 2. F; 3. F; 4. O; **Day 2:** 1–2. Answers will vary. **Day 3:** 1. B; 2. A; **Day 4:** 1. The following sentences should be crossed out: They are pretty. Aphids are bad. They fly fast.

Page 24

1. F; 2. O; 3. O; 4. F; 5. Answers will vary but should include factual information and details.
6. Answers will vary but should be opinions.

Page 25

Day 1: 1. Answers will vary.
Day 2: 1.

Day 3: 1. Answers will vary but may include the beach or a body of water. 2. Answers will vary but may include she put on her swimsuit, the sand, and the shells. 3. Answers will vary but may include that it is a hot day. Explanations will vary.
Day 4: 1. B; 2. A; 3. C

Page 26

1. the height of their plants; 2. Jason; 3. Thomas; 4. 12; 5. no; 6. shorter; 7. Answers will vary but may include that it takes time for plants to grow.

Page 27

Day 1: 1. true; 2. false; **Day 2:** 1. false; 2. true; **Day 3:** 1. B; 2. D **Day 4:** 1. C; 2. N—does not have four legs, fur, or whiskers; 3. N—does not have fur or whiskers

Page 28

1. 6; 2. 15; 3. 4; 4. 270; 5. 0; 6. both; 7. African elephant; 8. Asian elephant; 9. Answers will vary but may include seeing how many fingers are on its trunk, checking its diet or weight, or seeing whether it has tusks.

Page 29

Day 1: 1. All images should be circled.
Day 2: 1. mouse, cat, human, polar bear; 2. feather, candy bar, brick, car; **Day 3:** 1–4. Answers will vary. **Day 4:** 1. B; 2. C; 3. A; 4. D

Page 30

1. Answers will vary but may include hot—soup and fire; cold—ice, milk, and snow. 2. Answers will vary but should use terms to describe matter, such as size, shape, color, temperature, or texture. 3. Answers will vary. 4. short; 5. cold; 6. light; 7. smooth

Page 31

Day 1: 1. liquid; 2. solid; 3. gas; **Day 2:** 1. gas; 2. solid; 3. liquid; 4. solid; 5. liquid; **Day 3:** 1. water; 2. ice; 3. steam; **Day 4:** 1. Answers will vary but may include using a heat source.

Page 32

1. Check students' drawing and labeling.

Page 33

Day 1: 1. A; 2. C; 3. B; **Day 2:** 1–2. Answers will vary. **Day 3:** 1. Answers will vary but may include a container of water and various objects. **Day 4:** 1. 5, 2, 3, 1, 4

Page 34

1. 3; 2. 3; 3. 3; 4. 3; 5. 4; 6–8. Answers will vary. 9. Check students' drawing and labeling.

Page 35

Day 1: 1–2. Answers will vary. **Day 2:** 1. i, t (Light); 2. t (Heat); 3. o, u (Sound); **Day 3:** 1. to be wound up; 2. electricity; 3. batteries; 4. wind; **Day 4:** 1. B; 2. D

Page 36

1. C; 2. D; 3. B; 4. A; 5. F; 6. E; 7. Answers will vary but may include winding up, solar, batteries, or electricity. 8. Check students' drawing and labeling.

Page 37

Day 1: 1. sphere; 2. Answers will vary but may explain how the sphere is round while the other shapes have flat sides that would slow the object down. **Day 2:** 1. C; 2. A; 3. B; **Day 3:** 1. cheetah; 2. sphere; 3. empty wagon; **Day 4:** 1. C

Page 38

1. Check students' drawing and labeling.

Page 39

Day 1: 1. B; **Day 2:** 1. A. pull; B. push; C. pull; D. push; **Day 3:** 1–2. Check students' drawing. **Day 4:** 1. The ball would roll down the hill. 2. the force of gravity

Page 40

1–2. Answers will vary.

3.

Surface	Height of Bounce (in cm)
classroom floor	50
carpet	30
pillow	0

4. classroom floor; 5. pillow; 6. 20

Page 41

Day 1: 1. Answers will vary but might include that the pencil will fall to the ground. **Day 2:** 1. Gravity, Earth; 2. down; 3. moon; **Day 3:** 1. Answers will vary. 2. Answers will vary but may include helium balloons and astronauts in outer space. **Day 4:** 1. Check students' drawings. 2. Answers will vary but include describing that the pencil's weight needs to be evenly distributed.

Page 42

1. Answers will vary.

Page 43

Day 1: 1. Answers wil vary. 2. They are made of metal. **Day 2:** 1. D; 2. B; 3. C; 4. A; **Day 3:** 1. A; **Day 4:** 1. true; 2. true; 3. false

Page 44

1. metal door knob, nail clippers, blade of a scissors, cookie sheet; 2. Answers will vary but may include they are made of metal.

3.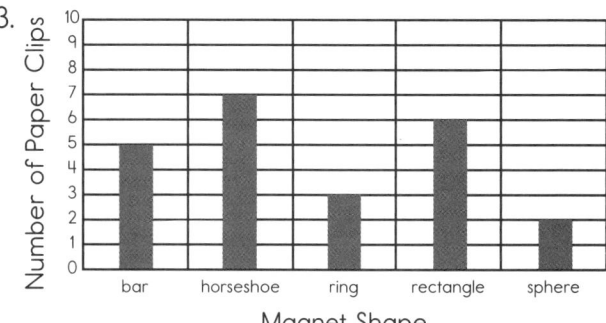

4. horseshoe; 5. sphere; 6. 4 paper clips

Page 45
Day 1: 1. P; 2. A; 3. A; 4. P; 5. P; 6. A; **Day 2:** 1. N; 2. L; 3. L; 4. L; 5. N; 6. N; **Day 3:** 1. Answers will vary but may include that it grows or reproduces. **Day 4:** 1. Living: tree, snake; Nonliving: rock, water; Pretend: unicorn, dragon

Page 46
1. Water; 2. cars; 3. Plants; 4. Elephants; 5. baby, fish, seal, tree, crab

Page 47
Day 1: 1. apple tree, grass, sunflower, lettuce; 2. Answers will vary. **Day 2:** 1. D, 2. C, 3. A, 4. B; **Day 3:** 1. apple, strawberries, pumpkin; 2. Answers will vary. **Day 4:** 1. A; 2. A; 3. D; 4. A; 5. D

Page 48
1. A. petal; B. leaf; C. stem; D. root; 2–3. Answers will vary.

Page 49
Day 1: 1. C; **Day 2:** 1. birds, turtles; 2. bats, birds; **Day 3:** 1. dog; 2. frog; 3. turtle; 4. penguin; 5. shark; **Day 4:** Answers will vary but may include the following: 1. slithers or crawls; 2. flies; 3. hops or jumps; 4. runs

Page 50
Answers will vary but may include the following: 1. cave; 2. nest or tree hollow; 3. hive; 4. house or apartment; 5. underground burrow; 6. dam

Page 51
Day 1: 1. Answers will vary but may include the sun is helpful as an energy source. 2. Answers will vary but may include the sun is harmful to the skin of animals. **Day 2:** 1. true; 2. true; 3. true; **Day 3:** 1. nests; 2. hide; 3. eat; **Day 4:** 1. soil; 2. plants; 3. leaves; 4. Insects

Page 52
1. Answers will vary but may include that sunlight helps plants to make food and grow. 2. Answers will vary but may include that animals use plants for food, shelter, and medicine. 3. Answers will vary but may include that animals pollinate plants, replant seeds, decompose dead plants, and provide carbon dioxide. 4. 6; 5. 9; 6. 6

Page 53
Day 1: 1. B; 2. D; 3. A; 4. C; **Day 2:** 1. Answers will vary but may include in the grass or on a leaf. 2. Answers will vary but may include that the insect would not be able to be seen since it would be camouflaged. 3. Answers will vary but may include it needs to hide from predators. **Day 3:** 1. bee; 2. Flowers; 3. birds; 4. Skunks; **Day 4:** 1. meat; 2. lion; 3. plants; 4. cow

Page 54
1. A; 2. Check students' drawings. 3. hide; 4. beak; 5. hibernate; 6. roots

Page 55
Day 1: 1. A; 2. C; 3. B; **Day 2:** 1. Answers will vary. **Day 3:** 1. Answers might include height, freckles, hair color, or eye color. **Day 4:** 1. shape of leaves, the seeds all grew into sunflowers

Page 56
1. Answers may include hair color, eye color, height, or interests. 2. Answers will vary but may include they are all mammals and have body systems. 3. 5; 4. 2; 5. 3; 6. 2; 7. a straight tail; 8. black fur, a long coat, and a straight tail; 9. Answers will vary but may include that the puppies got some traits from the mother and other traits from the father.

Page 57
Day 1: 1. Answers will vary but may include thick fur, blubber, or a bushy tail to wrap around the animal's body. **Day 2:** 1. tundra; 2. desert; 3. rain forest; **Day 3:** 1. C; 2. A; 3. B; **Day 4:** 1. 4; 2. 8; 3. 12

Page 58

1. C; 2. E; 3. B; 4. A; 5. D; 6. Answers will vary.

Page 59

Day 1: 1. 2, 1, 5, 3, 4; **Day 2:** 1. 1, 4, 5, 3, 2;
Day 3: 1. seed, vine, flower, orange pumpkin;
Day 4: 1. Answers will vary but may include
learning how to walk and talk, and growing taller.

Page 60

1. a chick; 2. 3 weeks; 3. The arrow means that the
adult will lay eggs, and the cycle will start over
again. 4. 1 week

Page 61

Day 1: 1. skin; 2. Answers will vary but may include
that both protect the inside of the person or
tree. 3. Answers will vary. **Day 2:** 1. Answers will
vary but may include hair, nails, bones, and skin.
Day 3: 1. 14; 2. 28; 3. Answers will vary but include
adding the 2 bones in the thumb with the 3
bones in the other 4 fingers.
(2 + 3 + 3 + 3 + 3 = 14) Then, add another 14 for
the other hand. (14 + 14 = 28) **Day 4:** 1. bones;
2. muscles; 3. lungs; 4. heart

Page 62

1. 32 > 30; 2. 32 < 42; 3. pigs; 4. A. nose; B. eye;
C. elbow; D. hand; E. knee; F. foot

Page 63

Day 1: 1. nighttime; 2. noon; 3. Answers will vary
but may include bats, raccoons, skunks, or moths.
4. sleep; **Day 2:** 1. night; 2. day; 3. day; 4. night;
5. day; **Day 3:** 1. F; 2. F; 3. T; 4. T; **Day 4:** 1. Things
I Can See During the Day: moon, sun, butterfly,
owl, cardinal; Things I Can See at Night: owl,
moon, stars, bat

Page 64

1. C; 2. B; 3–4. Check students' drawing
and labeling.

Page 65

Day 1: 1. Answers will vary but may include
mittens, coat, hat, scarf, long pants, or sweater.
2. Answers will vary but may include sandals,
shorts, a T-shirt, or a swimsuit. **Day 2:** 1. winter;
2. summer; 3. autumn; 4. spring;
Day 3: 1–4. Answers will vary. **Day 4:** 1. autumn;
2. summer; 3. summer; 4. autumn

Page 66

1. winter; 2. summer; 3. spring; 4. autumn;
5–8. Answers will vary but may include 5. In
spring, flowers grow, it rains, and buds grow on
trees. 6. In summer, weather is hotter, there are
more insects outside, and trees are full of leaves.
7. In autumn, trees lose their leaves as the leaves
turn colors, and squirrels gather nuts. 8. In winter,
weather is colder, sometimes snow falls, and the
trees are bare.

Page 67

Day 1: 1. sun, skin; 2. rays; 3. sunblock;
Day 2: 1. 12; 2. 365; 3. It repeats and is the reason
for the seasonal cycle. **Day 3:** 1. i (light);
2. e, a (heat); 3. o, o (food); 4. o, e (computers);
Day 4: 1. false; 2. true; 3. false; 4. false; 5. true

Page 68

1. Answers will vary but may include, How long will
it take for the butter to melt in the sun?
2. Answers will vary. 3. Answers will vary but may
include a timer.

4.  5.

6. Answers will vary but may include looking at
where the tree's shadow falls.

Page 69
Day 1: 1. one; 2. Answers will vary but may include round, white, and crescent shaped. **Day 2:** 1. C; **Day 3:** 1. moon, Earth, sun; 2. Answers will vary. **Day 4:** 1. Answers will vary but should include that we see the moon at night because it reflects the light from the sun.

Page 70
1. A; 2. C; 3. B; 4. D; 5. moon; 6. moon; 7. sun; 8. both; 9. moon; 10. sun; 11. moon

Page 71
Day 1: 1. C; 2. A; **Day 2:** 1. Answers will vary. **Day 3:** 1. Friday; 2. Monday and Tuesday; 3. 14; **Day 4:** 1. 6; 2. Answers will vary but may include subtracting 14 from 20 or counting up from 14 to 20.

Page 72
1. 30; 2-5: Answers will vary but may include 2. sunny; 3. rainy; 4. snowy or cold; 5. windy

Page 73
Day 1: 1. All should be circled. 2. Answers will vary. **Day 2:** 1. true; 2. true; 3. false; **Day 3:** 1. a, e (water); 2. c (ice); 3. i (wind); **Day 4:** 1. Living: trees, birds, grass; Nonliving: water, rocks, minerals, soil

Page 74
1-6: Answers will vary. 7. roots; 8. Water; 9. soil; 10. tunnels; 11. Beavers

Page 75
Day 1: 1. Answers will vary but may include ocean, river, stream, lake, or pond. 2. Answers will vary. **Day 2:** 1. a, i (rain); 2. o, w (snow); 3. a, e (water); 4. e, a (streams); **Day 3:** 1. Ponds, streams, and lakes should be circled. 2. 7; **Day 4:** 1. water; 2. salty; 3. frozen; 4. groundwater

Page 76
1. Answers will vary but may include to drink, bathe, wash, or stay cool. 2. B; 3. C; 4. C; 5. puddle, pond, lake, ocean

Page 77
Day 1: 1. D; 2. C; 3. A; 4. B; **Day 2:** 1. Answers will vary but may include that it is easier to make changes on a computer, and it does not use paper. 2. Answers will vary but may include that you can transport paper and pencil easily, and you do not need electricity to write by hand. **Day 3:** 1. Answers will vary but may include to store data, create graphs, and do research. **Day 4:** 1. Answers will vary but may include not using the Internet unless an adult is present, not eating or drinking near the computer, and not clicking on pop-up windows or ads.

Page 78
1. save; 2. scroll; 3. print; 4. font; 5. Check students' letters.

Page 79
Day 1: 1. Answers will vary but may include a hammer, a saw, nails, screws, and glue. 2. Answers will vary but may include paper and scissors. 3. Answers will vary but may include that you need different tools to work with different materials. **Day 2:** 1-2. Answers will vary. **Day 3:** 1. Answers will vary. **Day 4:** 1-3. Answers will vary.

Page 80
1-4. Answers will vary.

Page 81

Day 1: 1. Answers will vary but may include eating healthy foods, exercising, and getting enough sleep. **Day 2:** 1. Fruits: banana, strawberries; Vegetables: carrot, lettuce; Grains: whole wheat roll, rice; Protein: walnuts, eggs; Dairy: cheese, milk; 2. 5; **Day 3:** 1. brush, floss, Wash, Throw away, Bathing; **Day 4:** 1. Answers will vary but may include after using the bathroom, before eating, after coughing or sneezing, after playing outside, and after touching a pet.

Page 82

1. Answers will vary but may include eating healthy foods, exercising, and getting enough sleep. Being healthy helps you fight disease and gives you more energy. 2–3. Answers will vary. 4. Check students' drawings.

Page 83

Day 1: 1. Answers will vary but may include for food, clothing, shelter, toys, or medicine, or to make paper for books. **Day 2:** 1. books, orange, tea; 2. butter, cheese, can of tuna; **Day 3:** 1. R; 2. N; 3. N; 4. R; 5. R; **Day 4:** 1–2. Answers will vary.

Page 84

1. Check students' writing. 2. Check students' drawings. 3. renewable; 4. Answers will vary but may include that new trees can be planted.

Page 85

Day 1: 1. P, A, N; 2. P, A, N; 3. P, A, N; 4. N; **Day 2:** 1. 12; 2. 26; 3. 14; **Day 3:** 1. Trees are being cut down. 2. Answers will vary but may include animals losing their habitat or food source and that soil will not be protected from erosion. **Day 4:** 1. air; 2. water; 3. land

Page 86

1. Answers will vary but may include, they help it by conserving, reusing, or planting new trees. They harm it by polluting, cutting down trees, or using up natural resources. 2. Answers will vary.

Page 87

Day 1: 1. Answers will vary but may include paper, plastic, glass, and metal. 2. Answers will vary but may include food scraps, certain plastics, and foam takeout boxes. **Day 2:** 1. Answers will vary but may include planting new trees. **Day 3:** 1. reusing, 2. reducing, 3. recycling; **Day 4:** 1. Answers will vary but may include turning off lights when not in use, turning off faucets when not in use, and taking short showers. 2. Answers will vary but may include using a reusable water bottle or lunch bag and using the backs of paper. 3. Answers will vary but may include composting or using rain barrels to collect water.

Page 88

1. Can Be Recycled: soda can, glass bottle, newspaper; Cannot Be Recycled: banana peel, foam takeout box, slice of pizza; 2. Check students' writing.

Notes